CAREERS IN GEOPHYSICS

GEOPHYSICS IS THE STUDY OF THE earth's physical properties and processes acting upon, above, and within the earth. Geophysicists are scientists who apply the principles of physics, mathematics, chemistry, and engineering to study the earth and its environment. The studies are not limited to the earth's crust, but also include its internal composition, ground and surface waters, atmosphere, and various forces such as magnetism, electricity, and gravity.

Geophysicists use sophisticated equipment to take measurements of physical properties such as seismic waves, temperatures, natural electrical current, and gravity and magnetic fields. They collect data and samples in the field, and then work on computers, performing controlled experiments and analyzing samples.

Many geophysicists are involved in geophysical exploration. They work for companies that utilize natural resources, such as oil and minerals. The work involves surveying areas where they suspect resources may exist, then producing reports that their employers can use to determine if an investment in further exploration is worthwhile.

About 20 percent of geophysicists work for various local, state, and federal government agencies. Government work ranges from pure research, such as the development of global climate models, to monitoring how resource extraction is impacting the environment. Geophysicists also work for various private companies and consulting firms, environmental protection groups, and academic institutions. Some geophysicists are self-employed, working as independent consultants to assorted clients.

Within the major discipline of geophysics are numerous specialty fields, such as geodesy, seismology, atmospheric science, and oceanography. Geodesists look at the big picture, studying the planet's size, shape, gravitational field, tides, polar motion, and rotation. Seismologists endeavor to detect earthquakes and locate earthquake-related faults. Atmospheric science is devoted to the gaseous envelope surrounding the earth, including such phenomena as weather and climate. Some specialties are further broken down into different areas. Oceanographers, for example, generally study the world's oceans and coastal waters, but there are also physical oceanographers who focus solely on the relationships between the sea, weather, land, and climate. There are chemical oceanographers who are interested in the distribution of chemical compounds and chemical interactions within the ocean and on the sea floor. There are environmental oceanographers who investigate pollution problems and develop methods for more effective cleanup and prevention. The abundance of possible specialties has helped create an excellent job outlook.

In recent years, geophysicists have created and developed profound technical innovations. Some advances have essentially changed natural resources exploration and production. Many geophysicists are involved in hardware and software development. These professionals have pushed the boundaries of specialized computer applications, resulting in cutting-edge technology capable of processing huge 3-dimensional surveys.

Geophysics professionals work both indoors and outdoors. Some geophysicists are field oriented, some laboratory oriented, some theoretical, and some combine these areas. Some spend the majority of their time in a lab, but most divide their time between fieldwork and the lab. Work at remote sites is common. For example, a volcanologist will take field trips to take readings at active volcanoes, or an oceanographer may live on board an academic research ship for months at a time. Exploration geophysicists often work in foreign countries, sometimes under difficult conditions.

To become a geophysicist, you must have a curious mind, a fascination with natural phenomena, and an aptitude for math and science. A bachelor's degree is the basic educational requirement for an entry-level position, but most employers prefer a master's degree in geophysics. A PhD is necessary for most high-level research and university teaching positions.

The education required to enter this career will be challenging, but the effort will pay off. The starting pay is excellent, even for new graduates, and it gets better with time. The outlook for job growth is also good, with new jobs opening up at a faster pace than for most other occupations. There are opportunities for travel, too, both domestic and international. The work itself is very rewarding, according to most professionals in the field.

If you are imaginative, good at math and science, a critical thinker, comfortable with computers, and have a strong interest in what makes our planet tick, read on. A career in geophysics may be a good fit for you.

WHAT YOU CAN DO NOW

TO PREPARE FOR THE RIGOROUS college education that is required for a career in geophysics, it is best to start in high school. Take as many math and science classes as possible, especially any advanced courses that are available. Earth science, physics, and computer science are particularly important. Check the admission requirements for the colleges of your choice and make sure your curriculum includes every class you will need.

In addition to strong academic course work, employers like to see applicants with some first-hand experience in the field. This is not difficult to obtain, but the type of experience you have access to will depend on where you live. There are local field trips conducted by academic or government organizations just about everywhere that are related to geophysics. There are also student internships and summer programs open to high school students. The best opportunity available is with the US Geological Survey (USGS). Internships with the USGS are open to high school students and they even offer payment for the work performed.

Before setting out to become a geophysicist, first determine if the career is a good fit for you. Do you enjoy being outdoors? Are you comfortable with the idea of working in remote locations that may not have the modern conveniences you are used to? Geophysicists work both indoors and outdoors and need to be content stationed in many different environments. To learn more about whether you are well suited for this career, you should talk to people who are actually doing it. Ask your school counselor to invite some geophysicists to speak at your school on career days. Your counselor can also help you arrange job shadowing. Spending even a single day accompanying a geophysicist on the job will help you gain better understanding of what the job is all about.

HISTORY OF THE CAREER

GEOPHYSICS IS A RELATIVELY YOUNG science which only began to grow into a distinct professional discipline during the 19th century. However, the underlying phenomena – earthquakes, tsunamis, volcanic eruptions, and lightning – had been objects of fear and speculation since ancient times.

Long before it became a profession, the science was applied in numerous practical ways. For example, as long ago as the fourth century BC, hydrology was used to build complex irrigation works and large fresh water reservoirs in Sri Lanka. At the same time, the magnetic compass was being used in China for navigation on land. The compasses were not good for use on long sea voyages because the steel needles of the time could not retain their magnetism long enough to be useful. In 240 BC, Eratosthenes of Cyrene (an ancient Greek settlement near present-day Lybia) developed a system of latitude and longitude. A man ahead of his time, he deduced that the earth was round and measured the circumference of the earth using trigonometry and the angle of the Sun at more than one latitude in Egypt.

The modern science of geophysics was advanced by the publication *De Magnete*, written by William Gilbert in 1600. In the monograph, Gilbert asserted that compasses point north because the earth itself is magnetic – a fact that is common knowledge today, but considered mere theory at best in the 17th century. Gilbert backed up his claim through a series of meticulous experiments in magnetism, all of which were reported in the publication.

The 20th Century

The 20th century was a golden age for geophysics. Dramatic changes started early in the 1900s when hydrocarbons (petroleum and natural gas) were first used for generating power, heating, and transportation. For the first time, geophysical technology had a widespread practical use, guiding the extraction of hydrocarbons from the earth. Throughout the first half of the 20th century, there were three interrelated areas of progress in geophysics: oceanography, seismology, and magnetism.

Oceanography

There were several advances in physical oceanography. For example, acoustic measurements were first used to measure sea depth in 1914. That same year, Reginald Fessenden, a Canadian engineer, received a patent for an echo sounder. His invention is used for measuring the depth of water by sending pressure waves down to the bottom, then noting the time it took for the echo to return to the surface.

Seismology

Reginald Fessenden was also the first to use reflected seismic waves. The technology, which is similar to sonar and echolocation, is a method of exploration geophysics that can estimate the properties of the earth's subsurface using a controlled seismic source of energy. American geophysicist J. Clarence Karcher eventually commercialized the reflection seismograph. In 1917, he conducted the first reflection seismic experiment designed to locate oil. The experiment was so successful, it became the dominant method for the exploration of hydrocarbons.

Meanwhile, a German mine surveyor named Ludger Mintrop designed a mechanical seismograph that was used successfully to detect salt domes. Two decades later, Charles Richter and Beno Gutenberg developed the Richter Scale at the California Institute of Technology. Their invention is still used today to determine the size of earthquakes.

Magnetics

Prince Boris Golitsyn, a prominent Russian physicist is one of the founders of modern seismology due to his invention of the electromagnetic seismograph. The device worked by generating a signal from an object wrapped in wire, surrounded by a fixed magnet. The object moved in the magnet's magnetic field as the earth moved, producing a voltage that could be measured and recorded.

In the 1920s, Adolph Schmidt developed the first practical field magnetometer. Based on the magnetic needle system, the device was able to detect minerals in geological structures as well as oil exploration. Also in that decade, the aeromagnetic survey was first used. Now in common use, this method of geophysical survey towed a magnetometer behind an aircraft. The technique allowed for much larger areas of the earth's surface to be quickly covered.

In the second half of the 20th century, international scientific efforts covered important activity in all disciplines of geophysics: aurora and airglow, cosmic rays, geomagnetism, gravity, ionospheric physics, longitude and latitude determinations (precision mapping), meteorology, oceanography, seismology, and solar activity. Specific advances included:

Digital acquisition and processing revolutionized exploration geophysics.

3D seismic data acquisition and processing allowed geophysicists to visualize a site as a whole, rather than in the form of a flat elevation map. By the 1990s, 3D had evolved into 4D (time lapse 3D). 4D allows for more precise seismic imaging, allowing oil to be located more precisely.

Plate tectonics became the standard model of global geology. The movement of the plates slowly changes the size and shape below the earth's surface. These movements impact where oil and natural gas deposits can be found.

Global Position Systems (GPS) replaced traditional surveying methods in geophysical field surveys.

Radar imaging, developed in the late 1990s, made it possible to detect the presence, position, and motion of oil and gas deposits by analyzing the portion of the energy reflected from oil and gas deposits.

By the end of the 20th century, there had been many improvements in petroleum, mining, and groundwater geophysics. Among them, earthquake dangers had been minimized, making it safer for geophysicists to conduct soil/site investigations in earthquake prone areas.

Geophysics in the 21st Century

The most revolutionary advances in geophysics are continuing to be made in the 21st century. The stature of the profession is likewise growing with each new decade. For more than a century, geophysics has played a prominent role in discovering the evolution of the planetary surface, locating mineral and fossil fuel resources, and characterizing the history of the climate system. It has also defined the nature of hazards such as earthquakes, volcanoes, and tsunamis. A tremendous amount of progress has been made, yet much more is yet to be learned in this fascinating science. Geophysicists of the 21st century will be seeking answers to some of the greatest questions ever asked:

- How did the earth and other planets form?
- How does the earth's interior work, and how does it affect the surface?
- Why does the earth have plate tectonics and continents?
- What causes climate to change, how much can it change, and can it be changed according to man's actions?
- Can earthquakes and volcanic eruptions be predicted in time to avoid the consequences?

The answers to these questions will not come quickly or easily. It will take sustained and intense effort and the dedication of new generations of geophysicists focused on building upon current understanding and overcoming current limitations.

WHERE YOU WILL WORK

MOST GEOPHYSICS JOBS CAN BE divided into these categories: commercial, government, and nonprofit/academic.

The largest numbers of geophysicists in the private (commercial) sector are working in the exploration industries. This includes mostly petroleum and mining, though there are other types of work such as near-surface engineering, archaeological research, and groundwater studies. There are also companies involved in data processing related to natural resource exploration that employ large numbers of geophysicists.

In addition to oil and gas companies and mining concerns, employers of geophysicists include:

- Seismic data processing firms
- Well-logging companies
- Computer companies (software & hardware)
- Geophysical equipment companies
- Geological and geophysical consulting companies
- Research firms

- Integrated environmental firms
- Engineering firms
- Mapping service companies
- Instrument manufacturing and development firms

Geophysicists work at all levels of government from local agencies to branches of the military. The opportunities are quite varied, spanning all areas of geophysics from oceanography to atmospheric science.

The biggest employer of oceanographers and atmospheric scientists is the National Oceanic and Atmospheric Administration (NOAA). Those working for NOAA fill many different roles. Some work in forecasting offices while others conduct pure research or develop global climate models. Overall, the purpose of most of NOAA's work is to protect people and resources through early warning and detection of atmospheric conditions.

Another major federal employer is the National Aeronautics and Space Administration (NASA). Most of the geophysicists hired by NASA work at the Goddard Space Flight Center. The Planetary Geodynamics Laboratory within NSAS conducts research on the structure, dynamics, and evolution of the solid earth, moon, and planets. The laboratory is huge, divided into departments such as Geomagnetism, Topography and Surface Change, Crustal Deformation, Planetary Geology and Geophysics, Orbital-Rotational Interactions, and Space Geodesy.

State government agencies often hire geophysicists for geological surveys, but more often this type of work is done at the federal level. The Department of Interior's US Geological Survey (USGS) employs more than 10,000 geophysics scientists, technicians, and support staff. The National Geomagnetism Program alone has 14 observatories that monitor the magnetic field of the earth.

Geophysicists also work for nonprofit organizations that are involved in pure research, consulting, and resource exploration. Some become faculty members at universities, both public and private. Faculty members often have the opportunity to work on research projects conducted on campus.

Work Environment

Some geophysicists spend all day in an office working on computers processing data and modeling, or in a laboratory testing samples. Others spend 100 percent of their time in the field. Most often, however, it is a combination of both. All geophysicists have to do some work in the office, but they often welcome the chance to get out from behind the desk to venture out into the field. Field work may mean working in remote areas and living in bush camps, but not always. In addition to the rural environment, some field work is done in urban settings, collecting water or soil samples at problem sites, mapping, and conducting geophysical surveys. Oceanographers spend considerable time at sea on academic research ships. Wherever it is done, field work can mean being exposed to a variety of weather conditions and potentially dangerous situations.

Geophysics is a global endeavor. Professionals who like to travel will find plenty of opportunities to do so, both domestically and internationally. Many exploration geophysicists work in foreign countries, sometimes in remote areas and under difficult conditions. Some geophysicists travel to visit clients or collaborate with other scientists. Others live and work abroad, often moving from one country to another.

Work schedules vary. Geophysicists working in offices or laboratories usually work regular office hours, though the research conducted in some laboratories may require 24/7 shift work. Fieldwork, on the other hand is very unpredictable. Hours can be long and it is not unusual for geophysicists in the field to work evenings, weekends, and holidays.

THE WORK YOU WILL DO

THE TERM GEOPHYSICS LITERALLY means the physics of the earth. As a career field, geophysics deals with every aspect of the physical properties and processes of the earth, including earthquakes, winds, tides, volcanoes, gravity, magnetic and electrical fields, and weather. Geophysicists are interested in how these phenomena are formed and their effect on the earth.

Geophysicists are scientists who use physics, chemistry, geology, and advanced mathematics to study the earth. They explore the physical properties of the earth inside and out, from the ocean floor to the atmosphere. Without geophysics, our knowledge of the planet we call home would be minuscule.

Geophysicists use a variety of sophisticated instruments and research methods to:

- Measure changes in gravity and magnetic fields of the earth
- Detect radioactivity in rocks
- Find natural resources, such as ground water, oil, natural gas, potash, coal, iron, copper, and other minerals
- Assist with environmental cleanup and preservation
- Develop new techniques for earthquake monitoring and prediction
- Identify environmental hazards and evaluate areas for dams or construction sites

Daily duties vary depending on the type of job, but generally they involve collecting and examining samples from natural phenomena, studying readouts from measurement equipment, interpreting data that has been collected, and writing reports on the findings.

Most of the work done by geophysicists takes place in the lab, but in nearly every position some field work is required. This may be a normal, ongoing part of the job, such as exploring for various underground resources. It may be in response to an immediate phenomenon. For example, geophysicists would certainly rush to the site of an erupting volcano, movement of tectonic plates, or a major oil spill.

Geophysicists in research positions with the federal government or in universities have to come up with new ideas for research, develop those ideas into projects, and write grant proposals in order to fund the work. Unlike their salaried colleagues, geophysicists in consulting jobs need to market their skills and write proposals in order to have steady work.

Types of Geophysicists

Geophysics covers a broad range of earth science with dozens of different specialty areas. Although there are many more, most geophysicists specialize in petroleum, mining, environmental science, seismology, oceanography, and geomagnetism.

Petroleum Geophysicists

Petroleum geophysicists search the earth for new oil and gas deposits. They start at the computer, using 3-dimensional graphing software to analyze geological information and identify sites that should be explored. Potential sites may be on land or in ocean seabeds. They frequently use global positioning systems (GPS) and geographic information systems (GIS) to hone in on the locations of potential new reserves. Next, they gather rock and sediment samples from the sites by drilling a small sample well. The physical samples are then tested and evaluated using microscopes, geochemical analysis kits, and other laboratory equipment. In addition to determining the presence of hydrocarbons consistent with oil-rich areas, petroleum geophysicists are able to estimate the likely yield, the exact location of a crude oil deposit, and the depth to which companies should drill.

Mining Geophysicists

Mining geophysicists spend much of their time outdoors, using many different instruments to help them locate mineral resources. Copper, lead, zinc, iron, gold, potash – and even diamonds – are among these resources. Minerals have unique geophysical characteristics, some of which are only detectable through the use of highly sensitive instruments. Even then, mineral deposits can be elusive. It can take weeks or even months to collect enough precise data to correctly evaluate the economic potential of findings.

A new search will start with reconnaissance work, using instruments such as airborne magnetometers, various electromagnetic sensors, and radioactivity detectors. Information gathered from the airborne instruments is verified in detail with ground-based instruments, such as the Induced Potential, which is able to produce sharp 3-dimensional images. The I.P. sends electrical pulses into the ground, then measures the decay of voltage. Variation in the rate of decay indicates possible mineral deposits.

Environmental Geophysicists

Environmental geophysics is the use of geophysical methods to image and understand the properties and processes in the top 100 meters of the earth. This near-surface region has the most direct impact on our lives. These specialists spend an equal amount of time in the laboratory and out in the field, in both urban and rural areas. Through laboratory studies, theoretical modeling, and fieldwork, they are able to identify, map, and predict the presence and potential movement of water on or near the surface. In the same way, they can also identify contaminants in the soil within the upper 10 to 50 meters of the earth's surface.

As a practical application, this information is most useful in locating sites for safe underground waste disposal, finding existing problem sites that need to be cleaned up, and reducing the future impact on the environment. Environmental geophysicists are also called upon for more unusual activities, such as examining archaeological sites or assisting the police in their investigations of possible burial sites.

Seismologists

Seismology is the study of seismic waves and other vibrations in the earth. Seismic waves are waves of energy created when rock suddenly breaks apart within the earth, or by the slipping of tectonic plates. Most natural earthquakes are caused by sudden slippage along a fault line, but earthquakes can also occur as a result of volcano eruptions or nuclear bomb testing. Using data from seismographs and other geophysical instruments, these specialists are able to detect earthquakes and locate faults that could cause earthquakes and related phenomena, such as tsunamis.

Oceanographers

Oceanography is a specialty within environmental science. The oceans are vast, and the science covers a wide range of topics from plate tectonics and the geology of the sea floor, to the chemical and physical properties of the ocean – geological oceanography and physical oceanography.

Geological oceanographers study the features of the ocean floor, including its many rises and ridges, trenches and canyons, mountains and valleys. They are interested in how these features formed and how they are changing through erosional processes and seismic activity. Samples taken at the ocean floor offer a unique view of millions of years of history. There are also many undersea volcanoes to be studied, both live and dormant, that provide information on volcanic processes, hydrothermal circulation, magma genesis, and crustal formation.

Physical oceanographers study the conditions and processes within the ocean, such as temperature, salinity, density, tides and eddies, deep currents, surface waves, and tsunamis. They are interested in how these properties affect coastal areas in particular and the earth's climate in general.

Magnetic Geophysicists

Geomagnetics is concerned with the magnetic properties of earth. Specialists in this area use measurements of the earth's magnetic field, both current and from the past few centuries, to develop theoretical models that might explain the earth's origin.

A subset of this specialty, **paleomagnetism,** attempts to go much further back in history. Paleomagnetists are geophysicists who study the fossil magnetization in rocks and sediments from landmasses and oceans. They seek to understand the spreading of the sea floor, the wandering of the continents, and the many reversals of polarity that the earth's magnetic field has undergone throughout millions of years.

There are numerous other specialty areas for geophysicists such as geodesy, hydrology, paleontology, glaciology, gravity, and the atmospheric sciences.

Geodesists study the earth's size, shape, gravitational field, tides, polar motion, and rotation.

Hydrology is the study of the water cycle and water resources.

Paleontologists study fossils found in geological formations in order to trace the evolution of plant and animal life, and the geologic history of the earth.

Glaciology deals with snow or ice accumulation, which may be in the form of alpine and arctic glaciers, ice caps, ice sheets, and ice shelves. Glaciologists focus on the formation, movement, and effect of glaciers upon the earth's topography and the climate.

Gravity geophysicists use measurements of gravitational acceleration and gravitational potential at the earth's surface and above to look for mineral deposits. The surface gravitational field also provides information on the dynamics of tectonic plates.

Atmospheric geophysicists are concerned with climate change, atmospheric chemistry, radiative transfer, and other processes that affect the atmosphere. Some are involved in the subspecialty of atmospheric electricity, which involves studying lightning and the continual electrification in the air.

Career Advancement

Geophysicists often begin their careers in field exploration or as research assistants, lab analysts, or technicians. As they gain experience, they get assignments that are more difficult. Eventually, they are given supervisory responsibilities. Some are promoted to project leader, program manager, or to a senior research position. They may be charged with training new hires, and/or they may teach elective courses at local universities.

Geophysicists typically progress in their careers by branching out into new areas of study. These professionals enjoy learning how various systems interact with the ones they already know. This tends to lead to mastering new specialties, making them more useful to employers. Some write articles and books. Being published provides additional control over the direction of their research, which may mean leading research teams on important projects.

After 10 years in the profession, it is common for geophysicists to gravitate towards academia. Those who do not go into teaching may become managers or presidents of companies.

STORIES OF GEOPHYSICISTS

I Am a Freelance Geophysicist

"For me, every day is an adventure. Many geophysicists collect data in different parts of the world. They take measurements on land, at sea, in the air, and even from space. They work for a single employer who is focused on a particular activity, such as mining potash or studying the effects of climate change in certain regions. By contrast, I am a consultant, which means I am self-employed. In the course of a year, I might take on projects for more than a dozen organizations.

It's amazing how many different ways geophysics can be applied. I've hunted for unexploded bombs, scoured caves for minerals, measured the impact of oil spills, monitored underwater volcanoes, and even searched for pirate's treasure. I never know what I'm going to be doing next or where I'll be flying off to. Every single project is totally different. That's what makes my job so exciting.

The best part of being a geophysicist is the opportunity to travel. I get to see things I would never see as a tourist. Some things have never been seen by any human eyes before! I've been to every continent and beneath every ocean and sea. To be happy in this work, you must be willing and able to work outdoors in any kind of environment – and in any kind of weather. It can sometimes be stressful working in unfamiliar territory, and sometimes it's dangerous. It's never boring and the work makes me feel like I'm doing something useful. Plus, as a freelancer, I get paid more than my salaried colleagues. I get paid a day rate that is higher than most people get paid in a week.

Young scientists who want to be explorers can find an action-packed career in geophysics. You have to be prepared for long hours and time away from home, but if you want to see the world in a way that few
people ever will, there are plenty of opportunities awaiting you."

I Work on Ice Caps

"My job is to monitor the ice sheets in Antarctica, one of the most remote parts of the planet. The purpose of my work is to find out how the ice sheets are responding to global warming. My team is looking at whether the world's biggest ice caps melting are causing sea levels to rise, and if so, how fast is it happening. The speed is of particular importance to people who live in towns that would be flooded.

Like most geophysicists, I spend a lot of time on a computer. In addition to calculations, I use the computer to write up my reports and distribute them. Out in the field, I use standard geophysics techniques such as radar and seismic measurements. That involves using explosives to send sound waves into glaciers. I think for any geophysicist, fieldwork is the fun part. I love being out on the ice, even though I have to live in a tent and melt ice for drinking water. This isn't for everyone though. It requires living in very harsh conditions and extreme cold. I have gone months without being able to shower and a typical meal consists of dried food. But if you can cope with the conditions, there are tons of opportunities and the pay is excellent."

I Study Craters

"There are different kinds of craters, internal and external, or impact and volcanic craters. Some are on the surface and others lie beneath sediment. I look at the ways earth erosion affects existing craters and vice versa, and how volcanic craters have formed some of earth's most distinctive basins and lakes. My research findings are continually added to a large database that identifies every crater on the planet's surface.

I mostly do research, working on a computer. Very often, I'll work from home. But I don't want to spend my life sitting around in a lab coat, so I also participate in exploration projects whenever they come up. That means I join a team that goes out and searches for undiscovered craters. I also get out from behind the computer to attend conferences. There are dozens of these every year, that are both very specific to my field or more general. I go to present my work and also to compare the work I do with other scientists. I also get to visit really fascinating places, like Amsterdam and Rome.

There are misconceptions about geophysicists. Probably the main one is that all we do is look for new oil deposits. While it is true that the oil industry employs more geophysicists than any other, there are plenty of other jobs open to geophysicists – most of which are far more interesting."

PERSONAL QUALITIES YOU NEED

ALL GEOPHYSICISTS ARE INTELLECTUALLY curious and are fascinated with the natural phenomena of the earth. They have a natural aptitude in science and mathematics. To build a successful career in geophysics, there are certain other personality traits and skills that are particularly important. Nurture these and you will not only perform better on the job, you will enjoy greater satisfaction from your work.

Communications skills

There is a common misconception that going into science means not having to write. However, writing is just as important in this field as being able to do research. If you cannot communicate well, how will anyone know that you just did the most ground-breaking research in the history of geophysics? Like most scientists, geophysicists write many reports and research papers. The purpose is usually to present their findings to clients, employers, and other professionals, many of whom do not have a background in any kind of science. Being able to present the results of scientific research in clear and concise language that anyone can understand requires excellent communications skills.

Interpersonal skills

Geophysics projects are rarely solo activities. Geophysicists usually work in teams with colleagues, engineers, technicians, and other scientists. Patience, a mature outlook, and a sense of professional duty will make you a valued team member. You should also enjoy sharing and debating complex ideas with others. If you prefer to work on your own with little input from others, this may not be the career for you.

Critical thinking skills

Geophysicists base their findings on sound observation and careful evaluation of data. They often work on complex projects filled with challenges and the answers may not come quickly or easily. Successful geophysicists thrive on these challenges. Strong problem solving skills are fundamental to scientific research of any kind. Solving problems should be exciting and feel natural to you if you are considering this career.

Outdoor skills

Geophysicists must be willing to work outdoors as well as indoors. Many spend nearly all of their time outdoors, often in remote locations without the benefit of modern amenities such as housing, running water, and central A/C. Camping skills can be extremely useful. Every geophysicist should know the basics – how to pitch a tent, gather and purify water, and build a fire. There are many other outdoor skills that could come in handy as well, such as being able to read a topographical map, handle a boat, scuba-dive, or pilot a small aircraft.

Physical fitness

Even if you love research and are perfectly happy sitting in a lab coat, poring over piles of data on a computer, you will need to go out into the field sometimes. You must be ready and willing to work in or out of the office and be able to withstand changing weather and other conditions outdoors. Fieldwork requires a decent level of physical fitness. Even in urban areas, there is testing and sampling equipment to carry. Fieldwork in rural and remote locations often involves hiking and other physical activities. Certain specialties of geophysics, such as environmental geophysics or seismology demand more outside work and physical fitness than others.

ATTRACTIVE FEATURES

A CAREER IN GEOPHYSICS IS appealing in many ways, not the least of which is excellent pay and benefits. Starting pay for beginners averages $50,000. More experienced geophysicists can earn more than $200,000 a year if they are self-employed consultants, work in the oil industry, or join the faculty of a prestigious university. The job outlook is good, too. Employment is expected to increase faster than for most other occupations throughout the coming decade.

A geophysics career can be what you make of it. The work can be focused on small details or the biggest threats to humankind. In any case, there is always something new to be working on. Many geophysicists say they feel like explorers, discovering things that no one else has ever seen before. Being the first person in the world to know about a new discovery, and knowing that other scientists will use your work to continue the journey of human progress can be very rewarding. The continuing challenge and opportunity to learn new things is the major reason geophysicists are so satisfied with their work.

The work never needs to be boring because there are so many career paths open to geophysicists. You can teach, work in a government or industry laboratory, work for a variety of private concerns, or provide consultation services. There are also numerous specialty areas you can focus on, such as oceanography, climatology, seismology, and planetary geophysics. There are opportunities to solve problems in archaeology, engineering, mineral and energy exploration, and the environment. You could even help solve crimes and hunt for buried treasure.

Travel is an option. There is no reason to get stuck in an underground laboratory if you do not want to. Geophysicists can live and work in many different countries. Some travel continually, going from one environment to another. Some locations will be challenging, like the site of an awakening volcano. Others will be fascinating, like the 10,000-year-old sand dunes that lie beneath the Irish Sea. Still others will be difficult and dirty. The fact is that every single project can land you in a different place, using different equipment, and looking for different things.

UNATTRACTIVE ASPECTS

THE DAILY REPETITIVE TASKS OF taking readings, assembling large bodies of data, and finding patterns in that data can become tedious at time. Although the goal of the research may be exciting, the scientific process can get monotonous.

Geophysicists often have to relocate for work. This could simply be another city in the US, or it could mean moving to another country. Sometimes they are assigned to remote places where readings can be taken without man-made equipment interfering. There may or may not be housing provided, and there almost certainly will not be electricity or running water. It is possible the only transportation in and out of the area will be by boat, horseback, or on foot. For those who love the outdoors, this can be acceptable, but for those who have been in the field for a number of years, it is not all that much fun. It can also be difficult being away from family and friends for extended periods.

Becoming a geophysicist requires a rigorous education. Although you can begin working after completing your bachelor's degree, most employers only want candidates with a master's degree. In order to teach at the university level, or have some say in the type of research you want to do, you will need a PhD. That means four years of college, plus five to seven years of graduate school – all of which is filled with demanding math and science courses.

EDUCATION AND TRAINING

A BACHELOR'S DEGREE IS THE MINIMUM required education in this field, although aspiring geophysicists should be aware that job opportunities are best for those with at least a master's degree. An undergraduate degree involves four years of study, with a major in geophysics, geoscience, or a combined geology and physics program. Students who are planning on continuing to graduate school should carefully consider their choice of major.

Geophysicists need an extensive background in sciences. Undergraduate courses will emphasize geology, physics, and mathematics. A basic core curriculum will include classes in chemistry, engineering, logic, stratigraphy, structural geography, and mineralogy along with basic physics subjects such as quantum mechanics, classical physics, electromagnetism, and gravity. Environmental and ecological sciences are also becoming increasingly important to employers.

It is also important to have skills in research and relevant computer programs. The more experience you can get with relevant computer platforms and software programs, the better. Geophysicists routinely work with digital mapping technologies, global positioning systems (GPS), data collection and analysis programs, and other forms of computer modeling.

Graduate School

Though an entry-level position such as research assistant may be attainable with a bachelor's degree, a master's degree in geophysics or a closely related field is highly recommended. It takes approximately two to three years to complete a master's degree program in geophysics. Senior level research and university teaching positions require a Ph.D.

Many universities offer master's and doctoral degrees in geophysics. They are not all the same, however. It is important that you take the time to do your homework and find the graduate school that best matches your interests and ambitions. This will depend largely on the interests of the professors at a particular school. You can find out what professors are interested in by looking at different university department webpages. Reading geophysical journals and noting the affiliations of authors of articles are other ways to find schools that align with your intended career path.

Career Experience

While a strong academic background in science and math is important, exposure to field work is preferred by most employers. Some companies expect more than a little experience – up to three years in some cases. Internships are often part of a geophysics college program. If that is not the case at your school, you should seek one out for yourself. Ask your professors or career counselor to help point you in the direction of internship or placement opportunities.

Your college education will give you a good grounding in theory, but every geophysics job is different. Even with some fieldwork under your belt, there will be specialized aspects to the work that you will need to learn about. For that reason, many companies provide in-house training programs for new hires. Initial programs typically last a few days to a month. Ongoing training will be provided by experienced geophysicists within the company. Hydroelectric power plants and research facilities conduct some of the most intense on-the-job training programs.

EARNINGS

Geophysicists are paid well. Starting salaries for new graduates range from $30,000 up to $50,000. Experienced professionals earn an average salary of about $100,000.

Earnings vary depending on skills, experience, and type of employer. Experience strongly influences income for this work. Geophysicists can expect to earn more the longer they stick with the job. The highest paying skills associated with this career are seismic interpretation and data modeling.

There are several types of employers: commercial, government, and academic. Within the commercial sector, the petroleum, mineral, and mining industries offer the highest salaries. Geophysicists working for these private concerns can expect bonuses and periodic pay raises. An experienced geophysicist in the oil industry could earn well in excess of $250,000 a year.

Both government and academic employment are highly competitive. Government jobs are considered attractive due to the prospect of long-term employment and excellent benefits, but government salaries tend to level off after five years. University professors, on the other hand, start out earning high salaries, and earnings increase continually after achieving tenure. Salaries do vary considerably from one institution to another and also depending on location. A new faculty member could earn anywhere from $60,000 to $100,000 a year, depending on location. A fully tenured faculty member could earn around $250,000 per year. That is enough to motivate some geophysicists who want high earnings, but are not interested in working for oil companies, to spend the extra few years it takes to earn a doctoral degree that is required for a faculty position at a university.

Geophysicists in all categories can expect benefits such as health insurance, paid vacations, retirement plans, and sick leave.

OPPORTUNITIES

THE NUMBER OF JOBS FOR GEOPHYSICISTS is predicted to grow about 10 percent over the next decade. That is about average job growth for most occupations. The primary drivers of this demand are the continuing need for energy, coupled with the trend towards more environmental protections and responsible land and resource management. Job prospects are expected to be especially good in the oil industry and at consulting firms.

The oil and gas industry has long been the leading employer of geophysicists. In fact, nearly 40 percent work in this field – a number that is expected to increase. In the past, this has been an industry with shaky job security. Employment tended to fluctuate with the price of oil. When prices dropped, producers cut back on exploration efforts and laid off many geophysicists in the process. When prices rose, exploration resumed and geophysicists were hired in large numbers. That boom or bust cycle has been tempered in recent years primarily through the adoption of new technologies that lower production costs dramatically. In addition, there is a growing worldwide demand for oil and gas, which has geophysicists exploring previously inaccessible sites.

Geophysicists who speak a foreign language and are willing to travel abroad are at an advantage. However, job seekers will have to be prepared to work in remote areas and in developing nations where conditions may not be ideal.

Two of the newer technologies being applied to the oil and gas industry are horizontal drilling and hydraulic fracturing (fracking). The success of these technologies has led to an increased demand for geophysicists. Extraction methods have led to a natural gas boom in the US, though not without controversy. Fracking, in particular, is a hotly debated environmental issue. This has had a positive effect on the job market for geophysicists, particularly those who specialize in environmental geophysics. More of those specialists are being employed to study the effects of fracking and horizontal drilling on the surrounding areas.

Environmental geophysicists, along with other specialists such as hydrologists, are needed to help develop alternative energy sources. It takes expert planning to construct wind farms, solar power plants, and geothermal power plants. While these energy sources are considered more desirable than offshore oil wells, they still have an affect on wildlife and other environmental factors. Geophysicists are needed to monitor the quality of the environment wherever alternative energies are being produced.

More and more geophysicists are working as consultants. In the private sector, corporations need technical assistance and environmental management plans. Employment is particularly strong in management, scientific, and technical consulting services. The trend in the public sector is also to give work to outside consultants rather than put more employees on government payrolls. Both state and federal government agencies are dealing with budget constraints. Even the US Geological Survey, a major employer of geophysicists, has shifted some of its hiring to consultants.

While it is still possible to land a job in geophysics with only a bachelor's degree, most college graduates will find that good opportunities for them are scarce. The opportunities for those with a master's degree are excellent, though, making it well worth the investment to spend a few more years in school.

GETTING STARTED

THE GEOPHYSICS UNIVERSE IS SMALL, which means the number of entry-level jobs may be limited. A good education, hands-on experience, and good contacts in the field are valuable assets. Together with dedication and persistence, you will be well prepared to get your new career in motion.

Networking is an important tool in the geophysics community. Make sure to build connections while you are still in school. Contacts of all kinds can provide job leads. Stay in touch with your professors, fellow students, and college counselors. Work-related contacts are the most valuable. Build relationships with the people you meet through internships, volunteer positions, and work-study programs. Meet additional geophysics professionals by attending trade shows, conferences, and seminars. You will find them listed on professional association websites and in trade journals.

The single best way to get your foot in the door is to be an intern.

Ideally, you will intern where you would like to work full time after graduation. Be prepared for some competition getting into the internship program of your choice. Make sure you meet application deadlines and get as many references as you can from your professors. Employers like to see several years of hands-on experience so you will need to participate in more than one internship program. In every program, make your time count. Go the extra mile and let your supervisors know you would like to work there full time in the future.

There are numerous volunteer and part-time job opportunities available for geophysics students. Summer camps, science museums, and schools all want to have college students around to speak with young, aspiring scientists. Many government agencies and nonprofit organizations, in particular, rely heavily on volunteers. If internships are not an option, volunteer positions can provide similar benefits such as meeting influential contacts that can help you reach career goals.

Beyond networking, there are other ways to find job opportunities. Join professional organizations. Look for an organization that is devoted to your particular field of interest. Your membership will give you access to job openings and invitations to local chapter meetings. Make a habit of reading the organization's newsletters and science journals regularly. They are a good resource for job postings. Your entry into a career in geophysics is likely to be a summer job or cooperative work program. Visit your school's career center regularly to ask about these and other opportunities. Ask to be included on career network emails that the school sends out. Sometimes recruiters come to colleges and universities. You do not want to miss out on making connections just because you did not know about the event in time. You can also use employment agencies that specialize in science careers.

When sending out résumés, let all of your professors know where you are applying. Since this is a small field, chances are that there are connections between your current professors and the employers you are contacting. One phone call or email from your professor to an old friend on the hiring committee can go a long way in making your application stand out. Direct connections like that can help you find an entry-level job without having to go through an employment agency.

ASSOCIATIONS

■ **Seismological Society of America**
http://www.seismosoc.org

■ **US National Committee for Geological Sciences**
http://sites.nationalacademies.org/PGA/biso/GS/index.htm

■ **American Geosciences Institute**
http://www.agiweb.org

■ **The International Union of Geological Sciences (IUGS)**
http://iugs.org

■ **The American Geophysical Union (AGU)**
https://sites.agu.org

■ **Society of Exploration Geophysicists (SEG)**
http://seg.org

■ **Association for the Sciences of Limnology and Oceanography (ASLO)**
http://aslo.org/index.php

WEBSITES

■ **US Geological Survey**
www.usgs.gov

■ **National Oceanic and Atmospheric Administration (NOAA)**
http://www.noaa.gov

■ **National Aeronautics and Space Administration (NASA)**
https://www.nasa.gov

Copyright 2017
Institute For Career Research
CAREERS INTERNET DATABASE

www.careers-internet.org

www.ingramcontent.com/pod-product-compliance
Lightning Source LLC
Chambersburg PA
CBHW061239180526
45170CB00003B/1373